Targeting Viral Protease Enzymes as a Treatment for HIV AIDS and COVID-19

By

Shefali Srivastava

An Honors Thesis submitted to the Department of Biological Sciences

in partial fulfillment of the requirements for

degree of Bachelor of Science

Meredith College

Raleigh, North Carolina

November 30, 2020

Honors Student _Shefali Srivastava_   **Date** 11/23/2020

Thesis Co-Director _Karthik Aghoram_   **Date** 11/23/2020

Thesis Co-Director _Walda Powell_   **Date** 11/23/2020

Honors Director _Jennifer D McMillen_   **Date** 11/23/2020

Publication Agreement

I hereby grant to Meredith College the non-exclusive right to reproduce, and/or distribute this work in whole or in part worldwide, in any format or medium for non-commercial, academic purposes only.

Readers of this work have the right to use it for non-commercial, academic purposes as defines by the "fair use" doctrine of U.S. copyright law, so long as all attributions and copyright statements are retained.

Meredith College may keep more than one copy of this submission for purposes of security, backup, and preservation.

Shefali Srivastava

November 30, 2020

Acknowledgements

I would like to take this opportunity to thank everyone who has helped me with this Honors Thesis:

- My family for their unconditional love and unwavering support throughout my research, and for encouraging me to challenge myself in this research project.

- The Department of Chemistry, Physics, and Geoscience as well as the Department of Biological Sciences for the opportunity to conduct research at Meredith College and providing lab rooms and equipment.

- Dr. Karthik Aghoram, for advising me on the direction of the research as well as guiding me during the Assay Testing phase of my research. Also, for helping me edit my Honors Thesis, and guiding the COVID-19 literary research.

- Dr. Walda Powell, for advising me on the direction of the research as well as guiding me during the synthesis phase of my HIV-AIDS research. Also, for helping me edit my Honors Thesis, and guiding the COVID-19 literary research.

- Dr. Cassandra Lilly, for overlooking my research on HIV AIDS, and for providing guidance through the synthetic and analytical aspects of my research.

- Ms. Jocelyn Towe, for assisting me during synthesis and analytical testing phases of my HIV-AIDS research.

- My colleagues, Ms. Kiley van Ryn and Ms. Reese Buie, for working with me on the COVID-19 literary research and for collaborating with me on the chapter pertaining to COVID-19 in my Honors Thesis.

Contents

Matrine Derivatives as Possible HIV Protease Inhibitors

Abstract

HIV/AIDS is caused by the Human Immunodeficiency Virus. Its life cycle consists of six steps: infection, reception at the cell, integration into the host DNA, production of non-functional polypeptide, cleavage of polypeptide, and rebudding from the affected cell. The cleavage of the polypeptide chain is catalyzed by the action of HIV protease - a possible target for enzyme inhibition. Oxymatrine is an alkaloid compound extracted from *Sophora flacescens*, a Chinese herb, which has been shown to increase cardiac function by reducing the risk of heart failure and cardiac fibrosis. Through previous research, this compound has also shown inhibition of the HIV protease enzyme, making it a possible treatment for HIV-AIDS. Currently, there is great demand to synthesize chemical derivatives of matrine to make compounds that are more hydrophilic for drug testing. The purpose of this study was to explore the inhibition properties of matrinic acid derivatives on the HIV Protease enzyme. Over the course of this study, two derivatives of matrinic acid were synthesized: methyl ester of matrine and matrinamide. They were then tested on an HIV Protease assay to see if they inhibit the enzyme. Neither of the compounds showed any activity for inhibition. While they were both soluble in water, matrine methyl ester proved to be an activator of the HIV Protease enzyme, instead of an inhibitor. However, more research is required to purify the compounds synthesized. Further research can be conducted, using the ester derivative of matrine to activate similar cysteine-aspartic proteases to facilitate apoptosis in cancerous cells.

Introduction

According to the World Health Organization (WHO), 76 million people have been infected with the Human Immunodeficiency Virus (HIV) and a staggering 33 million people have died of Acquired Immune Deficiency Syndrome (AIDS) caused by HIV[1]. Even after years of research, it remains a threat today: 38 million people globally had HIV at the end of 2019, with African regions remaining the most severely affected with nearly 3.7% of adults living with HIV[1].

There are very few treatments available to those inflicted with HIV. However, these treatments are safe and effective and have caused the life expectancy of HIV patients to increase dramatically. However, most HIV patients must take 3 or more drugs in what is called "combination therapy" or "antiretroviral therapy" to treat the disease[2]. Most treatments involve taking multiple pills daily, and frequently meeting with healthcare providers, and in a small number of people, these medications may also cause serious side effects. This treatment would then have to continue throughout the patients' lives since there is no cure for HIV infection or AIDS[1]. It is for these reasons that the National Institute of Allergy and Infectious Diseases (NIAID) supports research on HIV treatment to develop therapies that last longer and could be administered over larger intervals of time rather than daily[3]. These long acting drugs could take the form of injections, patches or implants. Moreover, the NIAID encourages research that aims to target stages of the HIV life cycle as this would be an effective long-lasting therapy[3].

There are several stages in the life cycle[4] of the HIV virus. The virus targets the CD4 receptor found on the surface of T-helper lymphocytes in humans and binds with this receptor. It then fuses with the host cell and releases viral RNA and enzymes into the cell. Viral enzyme reverse transcriptase converts the viral RNA into double stranded DNA. This viral DNA is then

2

able to enter the host cell's nucleus, where viral enzyme integrase integrates the HIV DNA into the host cell DNA. When the host cell becomes active, the host's enzyme RNA polymerase copies the embedded viral genomic material and produces mRNA as a blueprint to make viral proteins. Once the viral peptide chain is produced by the host cell, a viral enzyme called HIV Protease cleaves the long polypeptide chain into smaller, functional units that come together along with copies of the HIV RNA genetic material to form a new virus particle. The newly assembled virus buds out of the host cell and the cycle starts over again[4]. A possible target for the treatment of HIV is the cleavage of the long polypeptide chains, which is facilitated by the HIV Protease Enzyme. The inhibition of this enzyme would mean the life cycle of HIV virus is broken and the virus cannot replicate further and infect other cells. This kind of treatment would also have the benefit of being a potential long-lasting treatment as it targets the life cycle of the virus itself, slowing down the replication of the virus dramatically.

Oxymatrine is an alkaloid compound found naturally in the roots of the *Sophora flavascens* or the Ku shen plant, that has been used in the past as a Chinese herbal remedy for rheumatoid arthritis[5]. Previous research has also shown that oxymatrine has demonstrated anti-inflammatory, immune reaction inhibiting, antiviral activities in humans as well as demonstrating protection against chronic heart failure in rats[6]. Oxymatrine has also been shown to decrease ischemia[7] and prevent cardiac fibrosis[8] in rats. Through previous research at Meredith College in North Carolina, oxymatrine has also been established as an inhibitor of HIV Protease. Matrine, a compound structurally similar to oxymatrine, has been speculated to also be an inhibitor of HIV Protease enzyme. However, the goal of this research was to synthesize HIV Protease inhibiting compounds that are hydrophilic and more suitable for drug testing. Therefore, in this study, it was hypothesized that the hydrophilic derivatives of matrine, methyl ester of

matrine and matrinamide, are inhibitors of HIV Protease enzyme.

Synthesis and Methodology

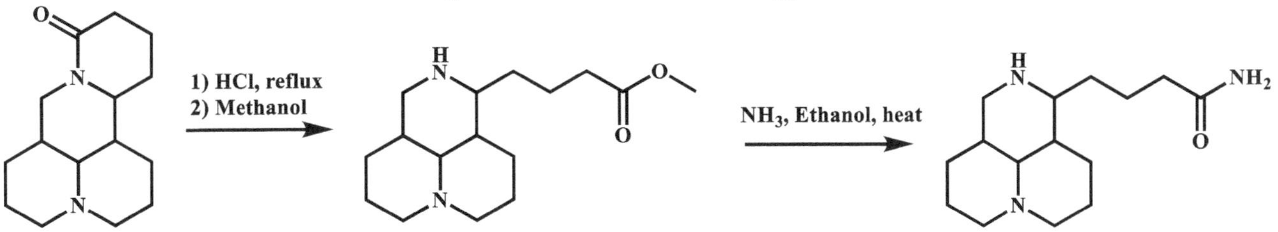

Matrine **Methyl ester of Matrine** **Amide derivative of Matrine**

Figure 1. Synthesis scheme for the preparation of derivatives of matrine.

The ester derivative of matrine was synthesized directly from matrine using hydrochloric acid and methanol. The ester was then used to make the amide derivative of matrine by the addition of the ammonia and ethanol over heat (Fig. 1).

Preparation of ester derivative of matrine (Fig. 2): The ester derivative of matrine was prepared by a ring-opening reaction involving matrine, hydrochloric acid, and methanol. 0.105g of matrine and 1 mL of 6M hydrochloric acid were refluxed for 6 hours. 1 mL of methanol was added to this solution and left to stir overnight. The solvent was then evaporated from this solution, to leave a glue-like residue. To get the solid residue, the solution was triturated with deuterated chloroform.

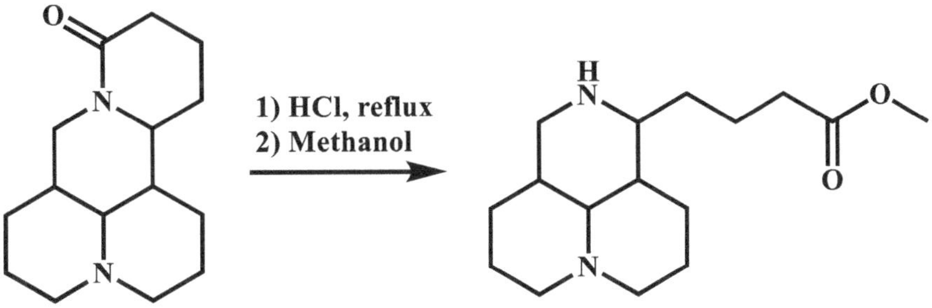

Matrine **Methyl ester of Matrine**

Figure 2. Synthesis of Methyl Ester of Matrine.

Preparation of amide derivative of matrine (Fig. 3): 0.35 g of the prepared ester derivative was stirred and refluxed overnight with 2 mL of methanol, 0.02 g of calcium chloride, and 1.5

mL of ammonia. The solution was then cooled to room temperature, and the solvent was evaporated leaving behind the solid amide derivative of matrine. The amide derivative was dissolved in ethanol and vacuum-filtered to purify it.

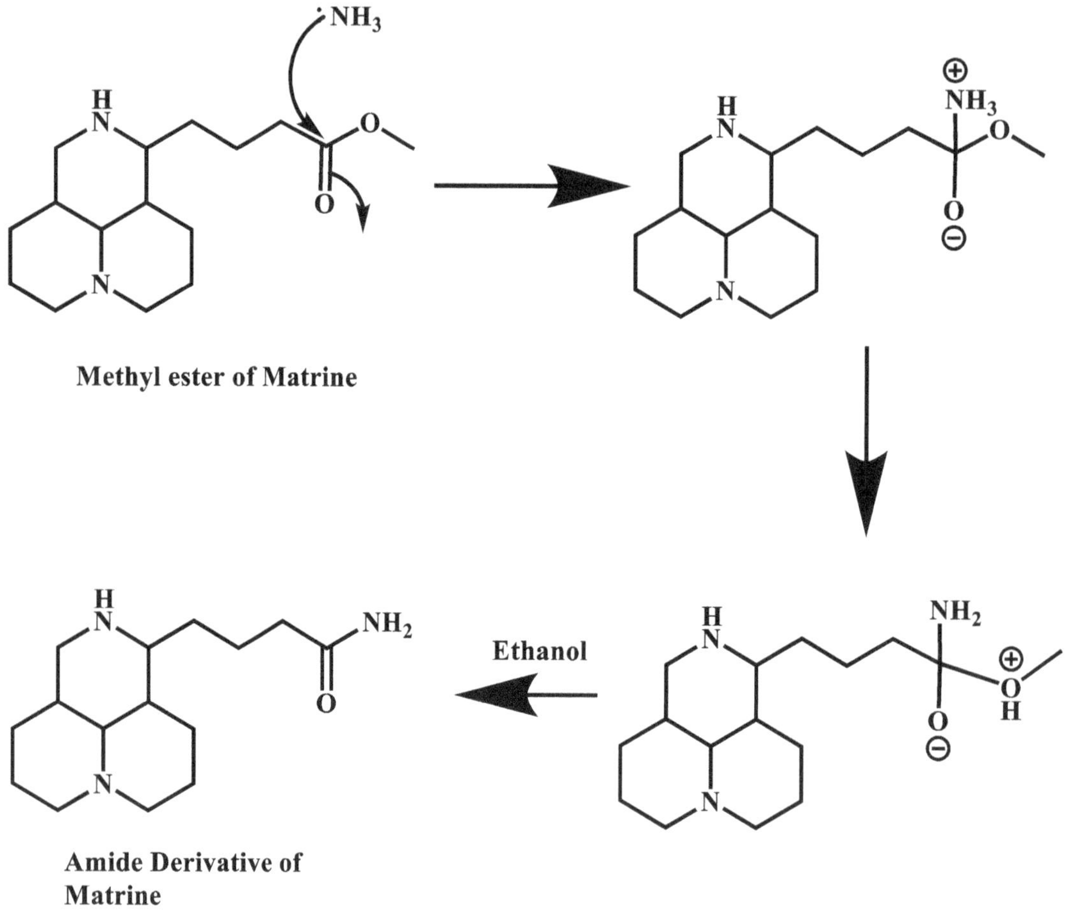

Methyl ester of Matrine

Amide Derivative of Matrine

Figure 3. Reaction steps in the synthesis of the Amide Derivative of Matrine

Analysis of Prepared Derivatives

Ester Derivative: the IR spectrum has peaks at 3389, 2939, 2736, 1723, 1433 cm^{-1}. The IR value of 1723 cm^{-1} is from the carbon-oxygen double bond in the ester. Mass Spectrometry resulted in peaks at 281 (M), 248, 205, 177, 150, 137, 96, 41 m/z. The physical color of the residue was off-white. The residue was soluble in water. Instrumentation used: Cary 630 FTIR instrument from Agilent Technologies; Agilent 6850/5973N Gas Chromatograph/Mass Spectrometer.

Amide derivative: the IR spectrum has peaks at 3384, 2952, and 1600 cm^{-1}. The value of 3384 cm^{-1} is the N-H stretch. The HNMR spectrum (Deuterium oxide) showed peaks at 3.63 (s, 3H), 3.57 (m, 1H), 3.42 (m, 1H), 2.98 (m, 1H), 2.78 – 2.73 (m, 2H), 2.30 (m, 1H), 2.12 (s, 1H) cm^{-1}. The physical color of the residue was white. The residue was soluble in water. Instrumentation used: Cary 630 FTIR instrument from Agilent Technologies; Agilent 6850/5973N Gas Chromatograph/Mass Spectrometer; Varian XL-300 HNMR spectrometer.

Testing Inhibitory Effect of Ester Derivative of Matrine

HIV-Protease Assay Kit (ProAssay^{TM}) was used to test for inhibitory effect of the matrine derivatives on HIV Protease enzyme. The Assay Kit includes the buffer stock solution, the HIV Protease enzyme, and a suitable substrate that fluoresces when cleaved by the enzyme. In this way, the activity of the enzyme can be measured quantitatively using a fluorometer to measure the fluorescence of the solution.

A buffer stock solution was prepared for each experiment, and it contained 400 μL of HIV-1 PR buffer, 0.8 μL of 1 M DTT, and 400 μL of deionized water. A "Master Mix" stock solution was prepared for each experiment described below. This solution was prepared by mixing 800.8 μL of prepared buffer solution, 8 μL of HIV-1 PR substrate, 1.6 μL of HIV-1 Protease Enzyme. To prepare the synthesized test compounds, both the ester and amide derivatives were dissolved in water to achieve a concentration of 156 mM (stock concentration). The tests were conducted in dark environments to preserve the integrity of the light sensitive substrate. The wells were then covered with aluminum foil and left in a dark space for 2 hours. A fluorometer was then used to measure the fluorescence of the solutions in the wells to determine the inhibitory effects of the test compounds.

The first experiment (Experiment 1) that was conducted was to determine if either of the synthesized compounds (ester derivative and amide derivative of matrine) had any inhibitory effect on the HIV Protease enzyme. For this experiment, the fluorescence measurements of the wells with the synthesized compounds were compared to each other and to control variables (no inhibitor, known inhibitor, water, and empty well). Table 1 shows the detailed well configuration for this experiment.

Table 1. Well Configuration of Experiment 1.

Well number	Contents
1	100 µL Master Mix 4 µL Master Mix
2	100 µL Master Mix 4 µL known inhibitor (provided by the HIV-Protease Assay Kit)
3	100 µL Master Mix 4 µL Methyl Ester of Matrine (synthesized ester derivative of matrine)
4	100 µL Master Mix 4 µL Matrinamide (synthesized amide derivative of matrine)
5	100 µL Master Mix 4 µL Oxymatrinic Acid (secondary known inhibitor experimentally synthesized in a previous research project)
6	100 µL Master Mix 4 µL Water
7	Empty

A second experiment (Experiment 2) was conducted to determine if any fluorescence measurements were spurious measurements. This was done by making sure that the source of the fluorescence was legitimate. Like the first experiment, the fluorescence measurements of the synthesized compounds were compared to each other and to wells containing individual components of the well solutions. The measurements obtained would determine if any substance other than the product of the enzyme-substrate complex was providing spurious fluorescence. For this experiment, the buffer solution was first prepared (scaled down using 500 μL of HIV-1PR buffer, 0.8 μL of DTT, and 500 μL of deionized water). This solution is mixed with 10 μL of substrate and pipetted into wells 4, 5, and 6. Then, 1.0 μL of enzyme was added to the remaining buffer and substrate mix and pipetted into wells 1, 2, and 3. Table 2 shows the well configuration for this experiment.

Table 2. Well configuration for testing for spurious fluorescence (Experiment 2).

Well Number	Contents
1	100 μL of enzyme, substrate, and buffer mix
2	100 μL of enzyme, substrate, and buffer mix 4 μL of Matrinamide (synthesized amide derivative of matrine)
3	100 μL of enzyme, substrate, and buffer mix 4 μL of Methyl Ester of Matrine (synthesized ester derivative of matrine)
4	100 μL of buffer and substrate mix
5	100 μL of buffer and substrate mix 4 μL of Matrinamide (synthesized amide derivative of matrine)
6	100 μL buffer and substrate mix 4 μL of Methyl Ester of Matrine (synthesized ester derivative of matrine)
7	100 μL of buffer solution 0.8 μL of enzyme 4 μL of Matrinamide (synthesized amide derivative of matrine)
8	100 μL of buffer solution 0.8 μL of enzyme 4 μL Methyl Ester of Matrine (synthesized ester derivative of matrine)
9	100 μL of buffer solution 0.8 μL of enzyme

A final experiment (Experiment 3) was conducted to assess system error that may have occurred during previous experiments, in order to validate previous results. For this experiment, a serial dilution of the test compound was prepared and pipetted into wells containing the master

mix solution (buffer solution, DTT, enzyme, substrate). The fluorescence of these wells was

measured in comparison to each other and to that of wells containing control variables (known

inhibitors and water). Different concentrations were prepared using the stock concentration of

156mM of the ester derivative and water (Table 3). The well configuration for this experiment is

shown in Table 4.

Table 3. Preparation of different concentrations of Methyl ester of Matrine for Experiment 3.

Concentration of Ester Derivative	Preparation
156 mM (stock solution)	0.2051g of synthesized Methyl Ester of Matrine dissolved in 5.01 mL of water.
31.2 mM	10 μL of 156 mM solution mixed with 40 μL deionized water
6.24 mM	10 μL of 31.2 mM solution mixed with 40 μL deionized water
1.25 mM	10 μL of 6.24 mM solution mixed with 40 μL deionized water

Table 4. Well configuration for testing for system error (Experiment 3).

Well Number	Contents
1	6 mM Methyl ester of Matrine Master Mix
2	1.2 mM Methyl ester of Matrine Master Mix
3	0.24 mM Methyl ester of Matrine Master Mix
4	0.048 mM Methyl ester of Matrine Master Mix
5	6 mM Methyl ester of Matrine Master Mix 4 μL Oxymatrinic Acid
6	100 μL Master Mix 4 μL Oxymatrinic acid
7	100 μL Master Mix 4 μL known inhibitor (provided by the Assay Kit)
8	100 μL Master Mix and 4 μL water

Results

The results of the first experiment are shown in Figure 4. These results show that the well containing the master mix and the ester and amide derivatives had a higher fluorescence than the wells containing the master mix alone, and the master mix along with the known inhibitor and oxymatrinic acid. The fluorescence of the ester and amide wells was also greater than the fluorescence measurements of the empty well and the well containing the master mix with water.

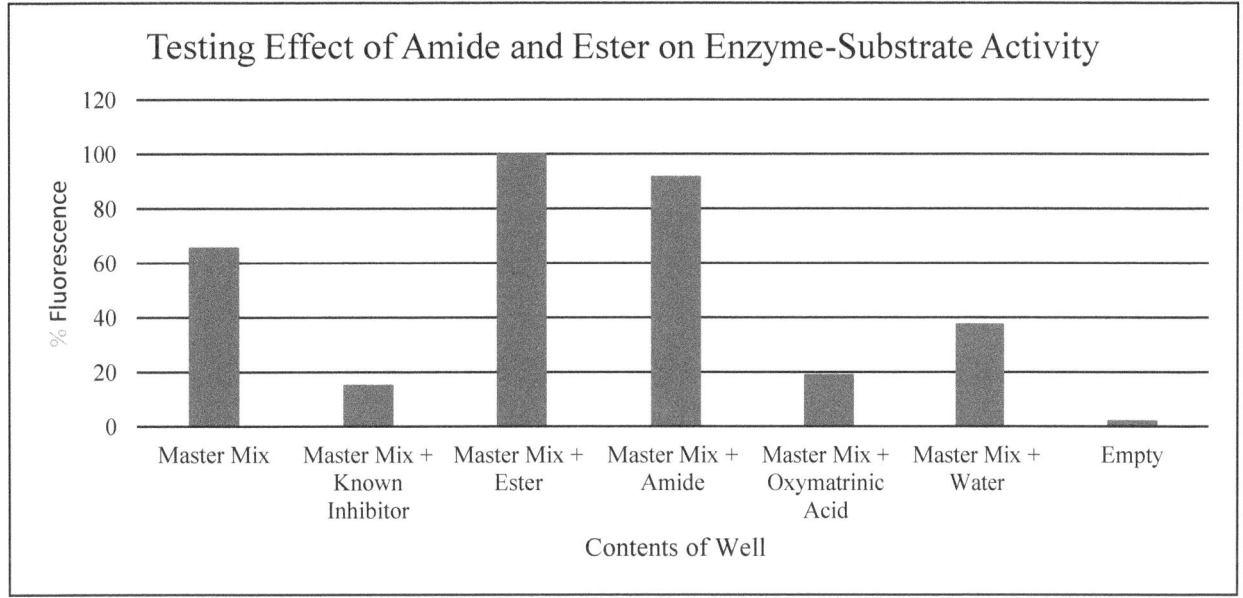

Figure 4. Bar graph showing the results of first experiment used to test the effect of amide and ester derivative on Enzyme-Substrate Activity. The "known inhibitor" was provided in the ProAssay Kit and Oxymatrinic acid was used as a control known inhibitor too.

Figure 6 shows the results for the second experiment. According to the results, very low florescence measurements were taken from the wells containing the amide and ester derivatives with the enzyme (and buffer) alone. The same is shown to be true for wells containing the amide and ester derivatives with the substrate (and buffer) alone. The only wells with noticeably higher fluorescence measurements were those which contain the ester and amide derivatives with the master mix containing both the enzyme and the substrate. These results also show that the well containing the ester and master mix had a 6.03% higher fluorescence measurement than the wells containing the amide and master mix, and the master mix alone.

11

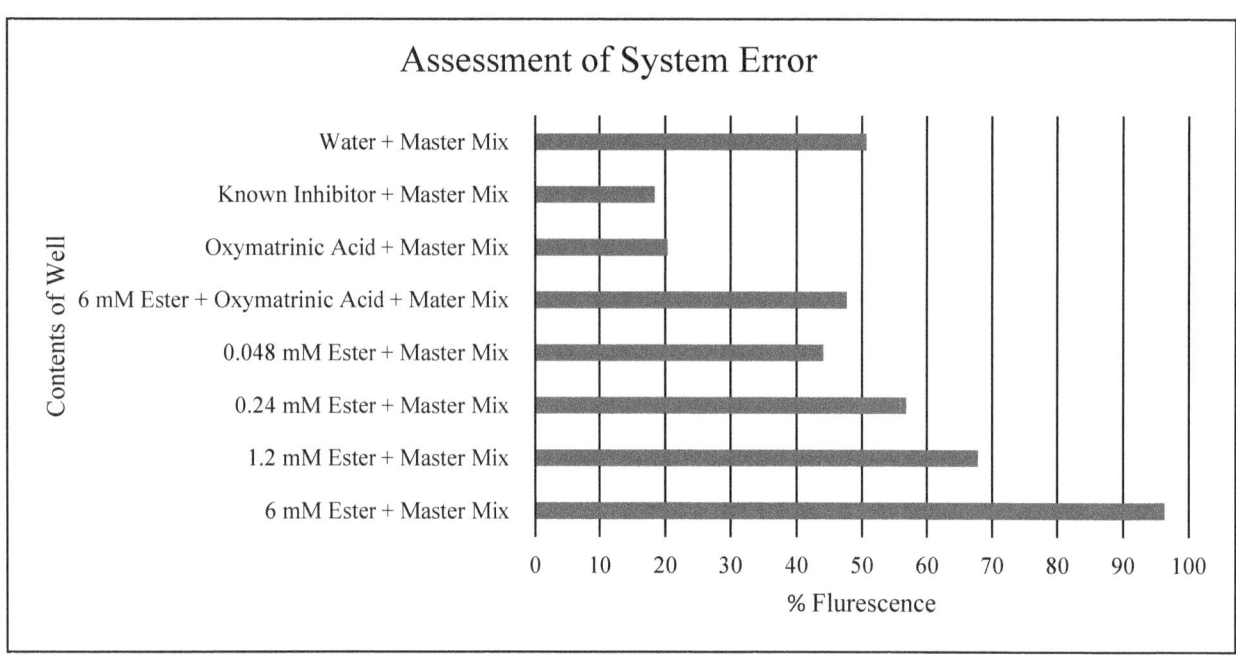

Figure 5. Bar graph showing the results from the assessment of system error (experiment 3) for the entire experiment, using a serial dilution of the ester.

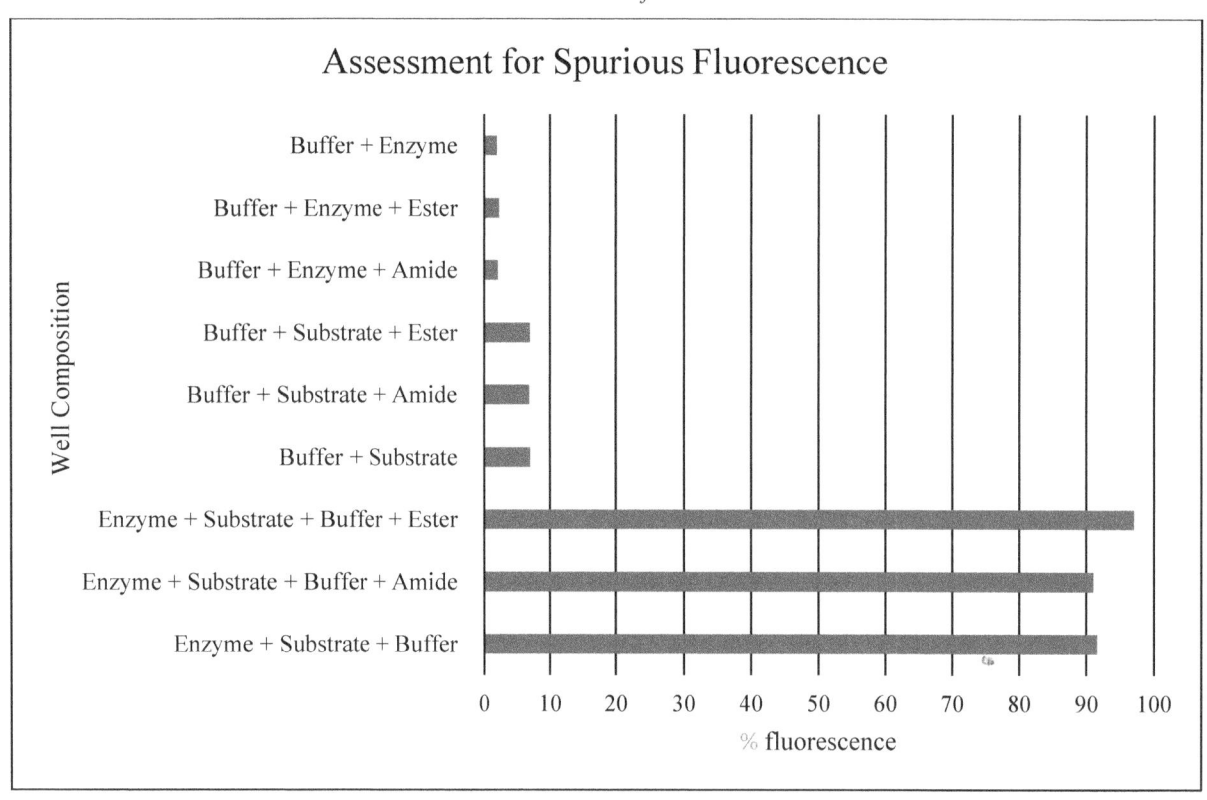

Figure 6. Bar graph showing the results for the experiment 2 conducted using test compounds (synthesized amide and ester derivatives of matrine) to test for spurious florescence.

According to Figure 5, the results from the assessment of system error showed that as the concentration of the ester in the well decreased, so did the % fluorescence measurements. The results also showed that the fluorescence measurement for the well containing the known inhibitor (provided by the ProAssay Kit) was considerably lower than the well containing just the master mix and water. The fluorescence measurement of the well containing the ester as well as the oxymatrinic acid (another known inhibitor) was higher than the fluorescence measurement of the well containing just the oxymatrinic acid and the master mix.

Discussion

From the results of Experiment 1, it can be deduced that the Methyl Ester of Matrine and the Matrinamide were both activators of the HIV Protease enzyme. However, since the assessment of system error was only performed using the Methyl Ester of Matrine, it can only be reliably concluded that the ester derivative of Matrine is an activator of HIV Protease enzyme. A second assessment of system error conducted for the Matrinamide would most likely yield a similar result. These results are quite surprising given that oxymatrine, a known inhibitor of HIV Protease Enzyme, is very similar to Matrine structurally. As is demonstrated in Figure 7., the only difference between Matrine and Oxymatrine is the extra oxygen on the Oxymatrine. This finding suggests that this extra oxygen performs a critical function in the role of oxymatrine as an inhibitor of HIV Protease Enzyme.

The results from the second experiment (Fig. 6) were conducted to assess spurious fluorescence in the results. This experiment not only confirmed that the ester and amide were activators of the HIV Protease Enzyme, but it also confirmed that the synthesized ester and amide derivatives did not have any effect on the enzyme or substrate alone that would affect the fluorescence measurements. It also confirmed that the master mix (enzyme, substrate, and buffer) alone did not affect the fluorescence measurements. This experiment suggests that the only process that produced the fluorescence was the cleavage of the peptide bond by the HIV Protease enzyme.

The assessment of system error (Fig 5) showed that as the concentration of ester decreases in the well containing the master mix and ester, the fluorescence measurements decrease as well. this experiment implies that water had no effect on the fluorescence measurements and that the ester derivative had no effect on oxymatrinic acid that would result in

14

a false fluorescence measurement due to contamination. These results suggest that the higher

fluorescence measurements seen in Experiment 1 (Fig 4) are due to the activating effect of the

ester derivative on the enzyme substrate complex alone.

Even though the presence of an oxygen on Oxymatrine can explain the activating effect

of the synthesized compounds, their purity should also be considered. The infrared (IR) spectrum

of the ester showed a pattern of peaks that was similar to the one described by Wang et. al., but it

was not an exact match to their IR spectra for the same molecule. Moreover, the mass

spectroscopy analysis showed that the molecular ion peak for the synthesized ester for this study

was higher than what was recorded by Wang et. al. for the same molecule. The same can be said

about the synthesized amide derivative of Matrine: the IR spectra, MS, and HNMR data was

similar to but did not match the primary research source (Wang et. al.). Another aspect of the

quantitative analysis worth mentioning is that it was hard to find a suitable deprotonated solvent

for the Methyl Ester of Matrine: DMSO was a worthy candidate but it was near impossible to

evaporate the DMSO after the analysis in order to retrieve the synthesized ester. For this reason,

it was concluded that a comparative HNMR analysis of the ester derivative would not be

possible for this study.

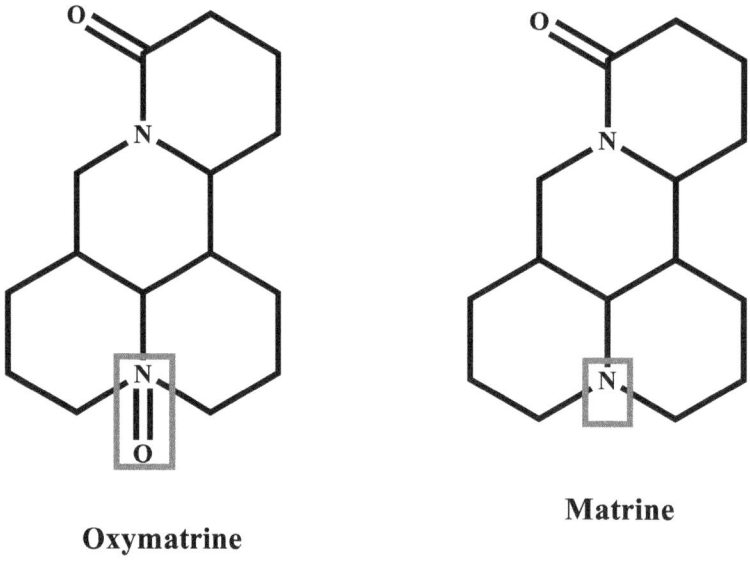

Oxymatrine

Matrine

Figure 7. Structural comparison of Oxymatrine and Matrine.

The experiments involving the HIV Protease Assay could also be discussed for the surprising results of this study. Though the ProAssay™ HIV-1 Protease Assay Kit has provided reliable and accurate data in previous studies, there were some irregularities in the amount of substrate provided in some of the kits used. Perhaps it should be considered that the substrate or enzyme used in the Assay itself was faulty. It is also important to note that due to the nationwide COVID-19 shutdown in 2020, multiple trials of synthesis, analysis, and assay testing could not be performed for the ester and amide derivatives, somewhat reducing the reliability of the results.

Conclusion and Further Research

The hypothesis that the ester and amide derivative of matrine would be inhibitors of HIV Protease Enzyme has been rejected given the data from the experiments conducted using the ProAssay™ HIV-1 Protease Assay Kit. The Methyl Ester of Matrine is seen to be an activator if HIV Protease. Though Matrinamide too showed activation of HIV Protease, more experiments need to be conducted to confirm this finding. This research calls into consideration the structural significance of the oxygen atom in the inhibitory effect of oxymatrine, as neither of the matrine derivatives showed inhibitory effects. Though the reliability of the data in this study can be improved, the results of this research opens several avenues for future studies to be conducted. The Methyl Ester of Matrine could be used as an activator of cysteine-aspartic proteases that perform a similar role of breaking peptide bonds to facilitate apoptosis in cancerous cells. The structural significance of oxygen in oxymatrine in the inhibition of HIV Protease may provide more knowledge about the structure of the HIV Protease enzyme and novel sites for inhibition. Similar ester and amide derivatives of oxymatrine could also be synthesized and tested for inhibition to get more insight into the structural significance of the oxygen on the oxymatrine. Statistical information about the function of oxymatrine as an inhibitor can be found to determine enzyme efficiency. More experiments should be conducted to ensure that the results of the fluorescence assay are valid and accurate. Perhaps, similar experiments should be conducted using different types of HIV Protease Assays from different companies. The ester and amide derivatives of matrine themselves could be synthesized using a different scheme in order to improve purity.

Identification and Synthesis of Potential Anti-SARS-CoV-2 Compounds

Written in collaboration with Reese Buie and Kiley Van Ryn

Abstract

The goal of this research was to identify and synthesize potential anti-SARS-CoV-2 compounds that can be used to break the lifecycle of the SARS-CoV-2 virus. As of October 24, 2020, there have been 230,000 deaths due to COVID 19 in the United States alone. The severity of the symptoms of COVID 19 varies, but these symptoms commonly include flu-like symptoms like fever, chills, cough, shortness of breath, fatigue, muscle or body aches, nausea, congestion, and may also include a new loss of sense of taste or smell, diarrhea. SARS-CoV-2 virus is composed of four proteins: The S, E, M, and N proteins. Once the virus has gained entry into the cell, it releases its viral genome. The genome is then translated by the viral polymerase protein following the replication and subgenomic RNA synthesis, 3CL^{pro} protease acts on the polypeptides to form functional S, E, and M proteins for the new virus. It is hypothesized that by inhibiting the HCoV-NL63 M^{pro}, and 3CL^{pro} viral protease enzymes, it would be possible to inhibit the replication cycle of the virus and prevent it from infecting more cells. This research outlines the potential of compounds to be inhibitors for HCoV-NL63 M^{pro} and 3CL^{pro} viral protease enzymes.

Introduction and Background

Severe Acute Respiratory Syndrome – Coronavirus 2 (SARS- CoV-2) is described as a complex RNA strand enveloped virus that causes coronavirus disease 2019 (COVID19). As of June 30th 2020, the global coronavirus pandemic has impacted over 10.4 million people worldwide and has led to over 507,000 deaths[9]. As of October 24, 2020, there have been 230,000 deaths due to COVID 19 in the United States alone[9]. Due to its long incubation period and high basic reproductive number, coronavirus disease 2019 (COVID 19) is particularly dangerous and the demand for new and effective treatment has skyrocketed[10]. This research aims to identify potential anti-SARS-CoV-2 compounds that prevent viral replication, and to propose synthesis schemes for the compounds that show significant promise.

The COVID 19 outbreak began in the first week of December of 2019 in the Wuhan province of China. It was not long before the causative virus was identified as SARS-CoV-2. After this initial outbreak, the virus had reportedly infected people who had visited or were employed at a meat market in Wuhan[11].

Current research pertaining to coronavirus speculates that they are RNA viruses distributed in a large number of animals[12]. Phylogenetic studies of coronaviruses suggest that α and β coronaviruses are particularly important to humans, as they have been reported to be zoonotic in nature[12]. In 2002, an unknown β-coronavirus SARS-CoV emerged in China and spread to several other countries, leading to over 800 deaths worldwide. Again, in 2012, another unknown β-coronavirus[12] was named – the Middle East Respiratory syndrome coronavirus (MERS-CoV) – and was reported to have a high case-fatality in human infections, with inefficient transmissibility between humans. In 2016, an α-coronavirus, thought to originate from bats, emerged in China and caused an epizoonotic disease in pigs called swine acute diarrhea syndrome coronavirus (SADS-CoV)[12].

Though it has been widely speculated that the SARS-CoV-2 virus originated in pangolins, phylogenetic research suggests otherwise. It is currently believed that the new SARS-CoV-2 virus underwent mutations that allowed it to infect humans and was most likely transmitted from bats[11]. Genetic sequencing studies on SARS-CoV-2 revealed 88% homology to two bat-derived CoV-like coronaviruses [bat-SL-CoVZC45 and bat-SL-CoVZXC21][13]. Furthermore, it is probable for bats to be the reservoir for SARS-CoV-2 due to similar zoonotic origins of other SARS-like coronaviruses.

Symptoms

The severity of the symptoms of COVID 19 varies, but these symptoms commonly include flu-like symptoms like fever, chills, cough, shortness of breath, fatigue, muscle or body aches, nausea, congestion, and may also include a new loss of sense of taste or smell, diarrhea[14]. The CDC recommends immediate medical attention if the person has trouble breathing, has persistent pain or pressure in the chest, new confusion, is unable to wake up or stay awake, or has bluish lips or face[14]. These symptoms may surface 2 to 14 days after infection, and it is unclear whether the person infected with the virus is able to transmit the disease while being asymptomatic[14].

Structure of SARS-Cov-2 Virus and Targeting the Life Cycle

SARS-CoV-2 virus is composed of four proteins: The S, E, M, and N proteins. The S protein is the spike protein present on the outer surface of the virus. The E protein is the envelope around the virus, whereas the M protein constitutes the viral membrane and N protein makes up the nucleocapsids. The life cycle of the virus involves the binding of the virus to the host cell via the S protein on the virus. The virus then enters the cell through membrane fusion.

20

Once the virus has gained entry into the cell, it releases its viral genome. The genome is then translated by the viral polymerase protein following the replication and subgenomic RNA synthesis, 3CL^{pro} protease acts on the polypeptides to form functional S, E, and M proteins for the new virus. The viral genome encapsulated by the N protein will then bud into the membrane resulting in the formation of the mature virion, and then the release of this virion via exocytosis. At the end of the replication cycle, the newly released mature virion can further infect a new host cell. It is hypothesized that by inhibiting the HCoV-NL63 M^{pro}, and 3CL^{pro} viral protease enzymes, it would be possible to inhibit the replication cycle of the virus and prevent it from infecting more cells.

According to the Milken Institute, there are currently 233 treatments in consideration and 162 vaccines in development for the COVID-19 disease. Remdesivir (GS-5734) is the most promising drug that exhibits broad-spectrum antiviral activities against RNA viruses[15]. Since the R_0 value[10] of COVID-19 is 5.7, the process of producing vaccines must be hastened in order to protect as many people as possible as fast as possible.

Synthesizing and Testing Potential Inhibitors

Based on previous research showing evidence that oxymatrine inhibits viral proteases, like the HIV protease, it can be hypothesized that oxymatrine will inhibit the proteases found in the SARS-CoV-2 virus. An oxymatrinic ester would be more soluble in water and it is believed that the ester derivatives of certain peptidomimetic compounds were more successful inhibitors than others.

Peptidomimetic molecules form a covalent bond with the catalytic cysteine via nucleophilic attack on the catalytic dyad of 3CL^{pro} active site[16]. These kinds of inhibitors include Micheal acceptors, aldehydes, epoxy ketones, and halomethyl ketones[16]. It can therefore be hypothesized that peptidomimetic inhibitors would prevent the 3CL[pro] active site from binding to the substrate polypeptides, stopping the replication cycle.

The HCoV-NL63 M[pro] uniquely accepts the glutamine residue at the Nsp13/14 cleavage site[17] of the viral substrate polyprotein. It follows that a 5-membered ring derivative of glutamine could be a part of α-ketoamides as it would successfully mimic glutamine and enhance the power of inhibitors by up to 10-fold[17]. It can be hypothesized that this new rigid molecule would compete with the cleavage site and bind to the active site of the protease more easily as it would reduce the loss of entropy upon binding to the protease[17].

In order to test compounds as potential inhibitors, it is necessary to purchase protease assay kits and the chemical compounds that will potentially act as SARS-CoV-2 protease inhibitors. The 3CL Protease Assay Kit from BPS Bioscience[18] is designed to test the activity of inhibitor molecules on the chymotrypsin-like SARS-CoV-2 3CL Protease. This kit is based on the fluorescence of a polypeptide substrate when acted on by the protease enzyme, much like the HIV Protease Assay Kit[18]. However, this kit requires a fluorescent microplate reader capable of reading exc/em of 360nm/460nm. The kit come with recombinant 3CL protease, 3CL substrate, a buffer for the protease, DTT, a known inhibitor (to act as a positive control), a microliter plate, and sealing film[18]. The Papain-like Protease Assay Kit from BPS Bioscience is designed to test the activity of inhibitor molecules of the papain-like SARS-CoV-2 PLPro Protease. This kit is based on the fluorescence of an ubiquitinated substrate when acted on by the PLPro Protease[18]. This kit, too, requires a fluorescent microplate reader capable of reading exc/em of 360nm/460nm. The kit comes with recombinant papain-like protease, ubiquitinated substrate, assay buffer, DTT, a known inhibitor, a microplate liter, and a sealing film[18]. As an alternative, SARS-CoV-2 Papain-like Protease Protein[19] may be purchased separately instead. This protein is delivered as a bulk protein in a solution of HEPES, NaCl, TCEP, and glycerol at a pH of 7.5. The papain-like protease in coronavirus operates on no less than 11 cleavage sites on the large polyprotein, and the inhibition of this enzyme will prevent virus replication.

The hydrolase Assay Kit can also provide evidence for the function of a desired protein and allows qualification of the amount of enzyme present. This kit requires a UV Visible Spectrometer. Some of these compounds to be tested will have to be synthesized in the lab before being tested in the protease assay kit.

Synthesis of the ester of oxymatrine would involve reacting oxymatrine with hydrochloric acid in methanol in a reflux setup overnight. The solution would then be evaporated or triturated (if needed) to yield a methyl ester of oxymatrine. The reaction can be performed with ethanol instead to yield an ethyl ester.

Another way to produce the ester derivative of oxymatrine would be to react oxymatrine with potassium hydroxide in water being heated, the solution would then be refluxed for 12 hours. Hydrochloric acid would be added to the solution till it reaches a pH of 5.53. The resultant solution can then be evaporated to give the solid carboxylic acid derivative of oxymatrine. This product can then be reacted with thionyl chloride (on ice) and then ethanol while being refluxed overnight to give the ethyl ester of oxymatrine, the residue obtained can be reacted with dichloromethane and sodium bicarbonate to obtain a filtrate, which can be evaporated to give the solid form of the ethyl ester.

Bibliography

1. The Global HIV/AIDS Epidemic [Internet]. U.S. Department of Health and Human Services. [Updated July 7, 2020]. Available from: https://www.hiv.gov/hiv-basics/overview/data-and-trends/global-statistics

2. Belperio P, Comstock E, DeSilva K, Maguire E, McFarland L, Moanna A. How is HIV treated: Basics [Internet]. [place unknown]. The U.S. Department of Veteran Affairs. 2019. Available from: https://www.hiv.va.gov/patient/basics/HIVtreatment.asp

3. Future Directions for HIV Treatment Research [Internet]. Maryland (MD): National Institute of Allergy and Infectious Diseases. [Updated Aug 26, 2019; c2020]. Available from: https://www.niaid.nih.gov/diseases-conditions/future-hiv-treatment

4. The HIV Life Cycle [Internet]. Maryland (MD): AIDSinfo – Service of the U.S. Department of Health and Human Services. Available from: https://med.unr.edu/Documents/unsom/statewide/aidsetc/HIVLifeCycle_FS_en.pdf

5. Aronson JK. Meyler's Side Effects of Drugs: The International Encyclopedia of Adverse Drug Reactions and Interactions. 16th Edition. Elsevier BV. Published 2016. Pages 229-236.

6. Hu S, Tang Y, Shen Y, Ao H, Bai J, Wang Y, Yang Y. Protective Effect of Oxymatrine on Chronic Rat Heart Failure. J Physiol Sci. [location unknown] Published 2011. volume 61: Page 363-372.

7. Hong-li S, Li L, Shang L, Zhao D, Dong D, Qiao G, Liu Y, Chu W, Yang B. Cardioprotective Effects and Underlying Mechanisms of Oxymatrine Against Ischemic Myocardial Injuries of Rats. [location unknown] Phytotherapy Research. Published 2008. Volume 22: Pages 985-89.

8. Shen X, Yang Y, Xiao T, Peng J, Liu X. Protective Effect of Oxymatrine on Myocardial Fibrosis Induced by Acute Myocardial Infarction in Rats involved in TGF-b1-Smads Signal Pathway. [location unknown] Journal of Asian Natural Products Research. Published 2011. Volume 13: Pages 215-224.

9. Coronavirus Death Toll [Internet]. [location unknown] Worldometer [updated November 14, 2020]. Available from: https://www.worldometers.info/coronavirus/coronavirus-death-toll/

10. Sanche A, Lin YT, Xu Chonggang, Romero-Severson E, Hengartner N, Ke R. High Contagiousness and Rapid Spread of Severe Acute Respiratory Syndrome Coronavirus 2. Emerg Infect Dis. 2020. Volume 26 (7): Pages 1470-1477. https://dx.doi.org/10.3201/eid2607.200282

11. Hasoksuz M, Kilic S, Sarac F. Coronaviruses and SARS-CoV-2. Turkish J Med Sci. 2020. Volume 50 (3): Pages 549-556.

12. Morens DM, Breman JG, Calisher CH, Doherty PC, Hahn BH, Keusch GT, Kramer LD, LeDuc JW, Monath TP, Taubenberger JK. The Origin of COVID-19 and Why It Matters. Am J Trop Med Hyg. 2020. Volume 103 (3): Pages 955-959. Doi: 10.4269/ajtmh.20-0849.

13. Malik YA. Properties of Coronavirus and SARS-CoV-2. Malays J Pathol. 2020. Volume 42 (1): Pages 3-11.

14. Coronavirus Disease 2019: Symptoms [Internet]. Center for Disease Control and Prevention. [updated 2020]. Available from: https://www.cdc.gov/coronavirus/2019-ncov/symptoms-testing/symptoms.html

15. Tu YF, Chien C-S, Yarmishyn AA, Lin YY, Luo YH, Lin YT, Lai WY, Yang DM, Chou SJ, Yang YP, Wang ML, Chiou SH. Int. J. Mol. Sci. 2020. Volume 21 (7): Page 2657. https://doi.org/10.3390/ijms21072657

16. Pillaiyer T, Manicham M, Namasivayam V, Hayashi Y, Jung S. An Overview of Severe Acute Respiratory Syndrome-Coronavirus (SARS-CoV) 3CL Protease Inhibitors: Peptidomimetics and Small Molecule Chemotherapy. J Med Chem. 2016. Volume 59: 6595-6628.

17. Zhang L, Lin D, Kusov Y, Nian Y, Ma Q, Wang J, von Brunn A, Leyssen P, Lanko K, Neyts J, de Wilde A, Snijder E, Liu H, Hilgenfeld R. α-Ketoamides as Broad Spectrum Inhibitors of Coronavirus and Enterovirus Replication: Structure-Based Design, Synthesis, and Activity Assessment. J Med Chem. 2020. Volume 63: Pages 4562-4578.

18. Coronavirus Assay Kits [Internet]. [location unkown]. BPS Bioscience: Scientist Founded, Scientist Driven. [Updated 2020]. Available from: https://bpsbioscience.com/research-areas/coronavirus/assay-kits

19. SARS-CoV-2 (COVID-19) Papain-like Protease Protein, His Tag [Internet]. Delaware (DE). Acro Biosystems. [Updated 2020]. Available from: https://www.acrobiosystems.com/P3165-SARS-CoV-2-%28COVID-19%29-Papain-like-Protease-Protein-His-Tag-%28active-enzyme%29.html

www.ingramcontent.com/pod-product-compliance
Lightning Source LLC
Chambersburg PA
CBHW080423190526
45161CB00004B/267